A Member of the International Code Family®

# International Code Council Electrical Code®

ADMINISTRATIVE PROVISIONS



2006

2006 ICC Electrical Code®—Administrative Provisions

First Printing: January 2006

ISBN-13: 978-1-58001-266-9 (soft)
ISBN-10: 1-58001-266-3 (soft)
ISBN-13: 978-1-58001-314-7 (e-document)
ISBN-10: 1-58001-314-7 (e-document)



PRINTED IN THE U.S.A

# PREFACE

## Introduction

Internationally, code officials recognize the need for provisions in administering the *National Electrical Code*® (NEC®). The ICC *Electrical Code®—Administrative Provisions*, in this 2006 edition, is designed to meet these needs through model code regulations. These administrative provisions correlate with provisions in the International Code system and are specifically geared to electrical code enforcement.

This 2006 edition is fully compatible with all the *International Codes*® (I-Codes®) published by the International Code Council (ICC)®, including the *International Building Code®, International Energy Conservation Code®, International Existing Building Code®, International Fire Code®, International Fuel Gas Code®, International Mechanical Code®,* ICC *Performance Code®, International Plumbing Code®, International Private Sewage Disposal Code®, International Property Maintenance Code®, International Residential Code®, International Wildland-Urban Interface Code*™ and *International Zoning Code®*.

The ICC *Electrical Code—Administrative Provisions* provides many benefits, among which is the model code development process that offers an international forum for electrical professionals to discuss performance and prescriptive code requirements. This forum provides an excellent arena to debate proposed revisions. This model code also encourages international consistency in the application of provisions.

## Development

The first edition of the ICC *Electrical Code—Administrative Provisions* (2000) was the culmination of an effort initiated in 1999 by a code development committee appointed by ICC and consisting of representatives of the three statutory members of the International Code Council at that time: Building Officials and Code Administrators International, Inc. (BOCA), International Conference of Building Officials (ICBO), and Southern Building Code Congress International (SBCCI). The intent was to draft a comprehensive set of regulations for administering the NEC. The Content of the administrative provisions of the *International Codes* was utilized as the basis for development of 2006 text. This 2006 edition presents the code as originally issued, with changes reflected in the 2003 edition, and further changes developed through the ICC Code Development Process through 2005. A new edition of the code is promulgated every three years.

## Adoption

The ICC *Electrical Code—Administrative Provisions* is available for adoption and use by jurisdictions internationally. Its use within a governmental jurisdiction is intended to be accomplished through adoption by reference in accordance with proceedings establishing the jurisdiction's laws. At the time of adoption, jurisdictions should insert the appropriate information in provisions requiring specific local information, such as the name of the adopting jurisdiction. These locations are shown in bracketed words in small capital letters in the code and in the sample ordinance. The sample adoption ordinance on page v addresses several key elements of a code adoption ordinance, including the information required for insertion into the code text.

## Maintenance

The ICC *Electrical Code—Administrative Provisions* is kept up to date through the review of proposed changes submitted by code enforcing officials, industry representatives, design professionals and other interested parties. Proposed changes are carefully considered through an open code development process in which all interested and affected parties may participate.

The contents of this work are subject to change both through the Code Development Cycles and the governmental body that enacts the code into law. For more information regarding the code development process, contact the Codes and Standards Development Department of the International Code Council.

While the development procedure of the ICC *Electrical Code—Administrative Provisions* assures the highest degree of care, ICC, its members and those participating in the development of this code do not accept any liability resulting from compliance or noncompliance with the provisions because the ICC does not have the power or authority to police or enforce compliance with the contents of this code. Only the governmental body that enacts the code into law has such authority.

## Letter Designations in Front of Section Numbers

In each code development cycle, proposed changes to the code are considered at the Code Development Hearings by the ICC Building Code Development Committee, whose action constitutes a recommendation to the voting membership for final action on the proposed changes. Proposed changes to a code section that has a number beginning with a letter in brackets are considered by a different code development committee. For example, proposed changes to code sections that have [F] in front of them (e.g., [F]1202.8) are considered by the International Fire Code Development Committee at the code development hearings.

The content of sections in this code that begin with a letter designation are maintained by another code development committee in accordance with the following:

[F] = International Fire Code Development Committee;

[FG] = International Fuel Gas Code Development Committee;

[M] = International Mechanical Code Development Committee; and

[PM] = International Property Maintenance Code Development Committee.

## Marginal Markings

Solid vertical lines in the margins within the body of the code indicate a technical change from the requirements of the 2003 edition. Deletion indicators in the form of an arrow ( ➡ ) are provided in the margin where an entire section, paragraph, exception or table has been deleted or an item in a list of items or a table has been deleted.

# ORDINANCE

The *International Codes* are designed and promulgated to be adopted by reference by ordinance. Jurisdictions wishing to adopt the 2006 ICC *Electrical Code—Administrative Provisions* as an enforceable regulation governing electrical systems should ensure that certain factual information is included in the adopting ordinance at the time adoption is being considered by the appropriate governmental body. The following sample adoption ordinance addresses several key elements of a code adoption ordinance, including the information required for insertion into the code text.

## SAMPLE ORDINANCE FOR ADOPTION OF THE ICC *ELECTRICAL CODE—ADMINISTRATIVE PROVISIONS* ORDINANCE NO.________

An ordinance of the **[JURISDICTION]** adopting the 2006 edition of the ICC *Electrical Code—Administrative Provisions*, regulating and governing the design, construction, quality of materials, erection, installation, alteration, repair, location, relocation, replacement, addition to, use or maintenance of electrical systems in the **[JURISDICTION]**; providing for the issuance of permits and collection of fees therefor; repealing Ordinance No. ______ of the **[JURISDICTION]** and all other ordinances and parts of the ordinances in conflict therewith.

The **[GOVERNING BODY]** of the **[JURISDICTION]** does ordain as follows:

**Section 1.** That a certain document, three (3) copies of which are on file in the office of the **[TITLE OF JURISDICTION'S KEEPER OF RECORDS]** of **[NAME OF JURISDICTION]**, being marked and designated as the ICC *Electrical Code—Administrative Provisions*, 2006 edition, as published by the International Code Council, be and is hereby adopted as the Electrical Code of the **[JURISDICTION]**, in the State of **[STATE NAME]** for regulating the design, construction, quality of materials, erection, installation, alteration, repair, location, relocation, replacement, addition to, use or maintenance of electrical systems as herein provided; providing for the issuance of permits and collection of fees therefor; and each and all of the regulations, provisions, penalties, conditions and terms of said Electrical Code on file in the office of the **[JURISDICTION]** are hereby referred to, adopted, and made a part hereof, as if fully set out in this ordinance, with the additions, insertions, deletions and changes, if any, prescribed in Section 2 of this ordinance.

**Section 2.** The following sections are hereby revised:

Section 101.1. Insert: **[NAME OF JURISDICTION]**

Section 404.2. Insert: **[APPROPRIATE SCHEDULE]**

**Section 3.** That Ordinance No. ______ of **[JURISDICTION]** entitled **[FILL IN HERE THE COMPLETE TITLE OF THE ORDINANCE OR ORDINANCES IN EFFECT AT THE PRESENT TIME SO THAT THEY WILL BE REPEALED BY DEFINITE MENTION]** and all other ordinances or parts of ordinances in conflict herewith are hereby repealed.

**Section 4.** That if any section, subsection, sentence, clause or phrase of this ordinance is, for any reason, held to be unconstitutional, such decision shall not affect the validity of the remaining portions of this ordinance. The **[GOVERNING BODY]** hereby declares that it would have passed this ordinance, and each section, subsection, clause or phrase thereof, irrespective of the fact that any one or more sections, subsections, sentences, clauses and phrases be declared unconstitutional.

**Section 5.** That nothing in this ordinance or in the Electrical Code hereby adopted shall be construed to affect any suit or proceeding impending in any court, or any rights acquired, or liability incurred, or any cause or causes of action acquired or existing, under any act or ordinance hereby repealed as cited in Section 3 of this ordinance; nor shall any just or legal right or remedy of any character be lost, impaired or affected by this ordinance.

**Section 6.** That the **[JURISDICTION'S KEEPER OF RECORDS]** is hereby ordered and directed to cause this ordinance to be published. (An additional provision may be required to direct the number of times the ordinance is to be published and to specify that it is to be in a newspaper in general circulation. Posting may also be required.)

**Section 7.** That this ordinance and the rules, regulations, provisions, requirements, orders and matters established and adopted hereby shall take effect and be in full force and effect **[TIME PERIOD]** from and after the date of its final passage and adoption.

# TABLE OF CONTENTS


**CHAPTER 1 SCOPE . . . . . . . . . . . . 1**

Section

101 General . . . . . . . . . . . . 1

102 Applicability . . . . . . . . . . . . 1

**CHAPTER 2 DEFINITIONS . . . . . . . . . . . . 3**

Section

201 General . . . . . . . . . . . . 3

202 General Definitions . . . . . . . . . . . . 3

**CHAPTER 3 ORGANIZATION AND ENFORCEMENT . . . . . . . . . . . . 5**

Section

301 Department of Electrical Inspection . . . . . . . . . . . . 5

302 Duties and Powers of the Code Official . . . . . . . . . . . . 5

303 Certificate of Occupancy . . . . . . . . . . . . 5

**CHAPTER 4 PERMITS AND FEES . . . . . . . . . . . . 7**

Section

401 General . . . . . . . . . . . . 7

402 Application . . . . . . . . . . . . 7

403 Conditions . . . . . . . . . . . . 7

404 Fees . . . . . . . . . . . . 8

**CHAPTER 5 CONSTRUCTION DOCUMENTS . . . 9**

Section

501 General . . . . . . . . . . . . 9

502 Examination of Documents . . . . . . . . . . . . 9

503 Design Professional in Responsible Charge . . . . . 9

504 Handling Submittals . . . . . . . . . . . . 10

**CHAPTER 6 APPROVAL . . . . . . . . . . . . 11**

Section

601 General . . . . . . . . . . . . 11

602 Testing . . . . . . . . . . . . 11

603 Alternative Engineered Design . . . . . . . . . . . . 11

**CHAPTER 7 INSPECTIONS AND TESTING . . . . 13**

Section

701 General . . . . . . . . . . . . 13

702 Required Inspections . . . . . . . . . . . . 13

703 Testing . . . . . . . . . . . . 14

**CHAPTER 8 SERVICE UTILITIES . . . . . . . . . . . . 15**

Section

801 General . . . . . . . . . . . . 15

**CHAPTER 9 UNSAFE SYSTEMS AND EQUIPMENT . . . . . . . . . . . . 17**

Section

901 Conditions . . . . . . . . . . . . 17

**CHAPTER 10 VIOLATIONS . . . . . . . . . . . . 19**

Section

1001 Unlawful Acts . . . . . . . . . . . . 19

1002 Notice of Violation . . . . . . . . . . . . 19

1003 Penalties . . . . . . . . . . . . 19

1004 Stop Work Order . . . . . . . . . . . . 19

**CHAPTER 11 MEANS OF APPEAL . . . . . . . . . . . . 21**

Section

1101 General . . . . . . . . . . . . 21

1102 Membership . . . . . . . . . . . . 21

1103 Procedures . . . . . . . . . . . . 21

**CHAPTER 12 ELECTRICAL PROVISIONS . . . . . . 23**

Section

1201 General . . . . . . . . . . . . 23

1202 Provisions . . . . . . . . . . . . 23

1203 Existing Electrical Facilities . . . . . . . . . . . . 24

**CHAPTER 13 REFERENCED STANDARDS . . . . . . 25**

**INDEX . . . . . . . . . . . . 27**

# CHAPTER 1
# SCOPE

## SECTION 101 GENERAL

**101.1 Title.** These regulations shall be known as the *Electrical Code–Administrative Provisions* of [NAME OF JURISDICTION] and shall be cited as such and will be referred to herein as "this code."

**101.2 Purpose.** The purpose of this code is to provide minimum standards to safeguard life or limb, health, property and public welfare by regulating and controlling the design, construction, installation, quality of materials, location, operation, and maintenance or use of electrical systems and equipment.

**101.3 Scope.** This code shall regulate the design, construction, installation, alteration, repairs, relocation, replacement, addition to, use or maintenance of electrical systems and equipment.

## SECTION 102 APPLICABILITY

**102.1 General.** The provisions of this code shall apply to all matters affecting or relating to structures and premises, as set forth in Section 101.

**102.1.1 Existing installations.** Except as otherwise provided for in this chapter, a provision in this code shall not require the removal, alteration or abandonment of, nor prevent the continued utilization and maintenance of, existing electrical systems and equipment lawfully in existence at the time of the adoption of this code.

**102.1.2 Maintenance.** Electrical systems, equipment, materials and appurtenances, both existing and new, and parts thereof shall be maintained in proper operating condition in accordance with the original design and in a safe, hazard-free condition. Devices or safeguards that are required by this code shall be maintained in compliance with the code edition under which installed. The owner or the owner's designated agent shall be responsible for the maintenance of the electrical systems and equipment. To determine compliance with this provision, the code official shall have the authority to require that the electrical systems and equipment be reinspected.

**102.1.3 Additions, alterations and repairs.** Additions, alterations, renovations and repairs to electrical systems and equipment shall conform to that required for new electrical systems and equipment without requiring that the existing electrical systems or equipment comply with all of the requirements of this code. Additions, alterations and repairs shall not cause existing electrical systems or equipment to become unsafe, hazardous or overloaded.

Minor additions, alterations, renovations and repairs to existing electrical systems and equipment shall meet the provisions for new construction, except where such work is performed in the same manner and arrangement as was in the existing system, is not hazardous and is approved.

**102.1.4 Change in occupancy.** It shall be unlawful to make a change in the occupancy of any structure that will subject the structure to any special provision of this code applicable to the new occupancy without approval. The code official shall certify that such structure meets the intent of the provisions of law governing building construction for the proposed new occupancy and that such change of occupancy does not result in any hazard to public health, safety or welfare.

**102.1.5 Moved buildings.** Electrical systems and equipment that are a part of buildings or structures moved into or within the jurisdiction shall comply with the provisions of this code for new installations.

**102.2 Differences.** Where, in any specific case, different sections of this code specify different materials, methods of construction or other requirements, the most restrictive shall govern. Where there is a conflict between a general requirement and a specific requirement, the specific requirement shall be applicable.

**102.3 Other laws.** The provisions of this code shall not be deemed to nullify any provisions of local, state or federal law.

**102.4 Validity.** In the event any part or provision of this code is held to be illegal or void, this shall not have the effect of making void or illegal any of the other parts or provisions thereof, which are determined to be legal; and it shall be presumed that this code would have been adopted without such illegal or invalid parts or provisions.

**102.4.1 Segregation of invalid provisions.** Any invalid part of this code shall be segregated from the remainder of this code by the court holding such part invalid, and the remainder shall remain effective.

**102.5 Application of references.** References to chapter or section numbers, or to provisions not specifically identified by number, shall be construed to refer to such chapters, sections or provisions of this code.

**102.6 Referenced codes and standards.** The codes and standards referenced in this code shall be those that are listed in Chapter 13 and such codes and standards shall be considered part of the requirements of this code to the prescribed extent of each such reference. Where differences occur between provisions of this code and referenced codes or standards, the provisions of this code shall apply.

**Exception:** Where enforcement of a code provision would violate the conditions of the listing of the equipment or appliance, the conditions of the listing and manufacturer's instructions shall apply.

**102.7 Appendices.** Provisions in the appendices shall not apply unless specifically referenced in the adopting ordinance.

**102.8 Subjects not regulated by this code.** Where no applicable standards or requirements are set forth in this code, or are contained within other laws, codes, regulations, ordinances or bylaws adopted by the jurisdiction, compliance with applicable standards of nationally recognized standards as are approved shall be deemed as prima facie evidence of compliance with the intent of this code. Nothing herein shall derogate from the authority of the code official to determine compliance with codes or standards for those activities or installations within the code official's jurisdiction or responsibility.

# CHAPTER 2
# DEFINITIONS

## SECTION 201
## GENERAL

**201.1 Scope.** Unless otherwise expressly stated, the following words and terms shall, for the purposes of this code, have the meanings indicated in this chapter.

**201.2 Interchangeability.** Words used in the present tense include the future; words in the masculine gender include the feminine and neuter; the singular number includes the plural and the plural, the singular.

**201.3 Terms defined in other codes.** Where terms are not defined in this code and are defined in the *International Building Code, International Energy Conservation Code, International Fire Code, International Fuel Gas Code, International Mechanical Code, International Plumbing Code, International Private Sewage Disposal Code, International Property Maintenance Code, International Residential Code, International Zoning Code* or NFPA 70, such terms shall have meanings ascribed to them as in those codes.

**201.4 Terms not defined.** Where terms are not defined through the methods authorized by this section, such terms shall have ordinarily accepted meanings such as the context implies.

## SECTION 202
## GENERAL DEFINITIONS

**APPROVED.** Approved by the code official or other authority having jurisdiction.

**APPROVED AGENCY.** An established and recognized agency regularly engaged in conducting tests or furnishing inspection services, where the agency has been approved by the code official.

**CODE OFFICIAL.** The officer or other designated authority charged with the administration and enforcement of this code, or a duly authorized representative.

**LISTED AND LISTING.** Equipment, appliances or materials included in a list published by a nationally recognized testing laboratory, inspection agency or other organization concerned with product evaluation that maintains periodic inspection of the production of listed equipment, appliances or materials, and whose listing states either that the equipment, appliances or materials meet nationally recognized standards, or has been tested and found suitable for use in a specified manner. Not all testing laboratories, inspection agencies and other organizations concerned with product evaluation use the same means for identifying listed equipment, appliances or materials. Some do not recognize equipment, appliances or materials as listed unless they are also labeled. The authority having jurisdiction shall utilize the system employed by the listing organization to identify a listed product.

**OCCUPANCY.** The purpose for which a building, or part thereof, is utilized or occupied.

# CHAPTER 3
# ORGANIZATION AND ENFORCEMENT

## SECTION 301
## DEPARTMENT OF ELECTRICAL INSPECTION

**301.1 Creation of enforcement agency.** The department of electrical inspection is hereby created and the official in charge thereof shall be known as the code official. The function of the department shall be to assist the code official in the administration and enforcement of the provisions of this code.

**301.2 Appointment.** The code official shall be appointed by the chief appointing authority of the jurisdiction.

**301.3 Deputies.** In accordance with the prescribed procedures of this jurisdiction and with the concurrence of the appointing authority, the code official shall have the authority to appoint a deputy code official, the related technical officers, inspectors, plans examiners and other employees. Such employees shall have powers as delegated by the code official.

## SECTION 302
## DUTIES AND POWERS OF THE CODE OFFICIAL

**302.1 General.** The code official is hereby authorized and directed to enforce the provisions of this code. The code official shall have the authority to render interpretations of this code, and to adopt policies, procedures, rules and regulations in order to clarify the application of its provisions. Such interpretations, policies, procedures, rules and regulations shall be in compliance with the intent and purpose of this code. Such policies and procedures shall not have the effect of waiving requirements specifically provided for in this code.

**302.2 Rule-making authority.** The code official shall have authority as necessary in the interest of public health, safety and general welfare, to adopt and promulgate rules and regulations and to designate requirements applicable because of local climatic or other conditions. Such rules shall not have the effect of waiving requirements specifically provided for in this code, or of violating accepted engineering methods involving public safety.

**302.3 Applications and permits.** The code official is authorized to receive applications, review construction documents and issue permits for the installation of electrical systems and equipment, inspect the premises for which such permits have been issued, and enforce compliance with the provisions of this code.

**302.4 Notices and orders.** The code official is authorized to issue all necessary notices or orders in accordance with Chapter 10 as are required to effect compliance with this code.

**302.5 Inspections.** The code official shall make all of the inspections necessary to determine compliance with the provisions of this code in accordance with Chapter 7.

**302.6 Identification.** The code official shall carry proper identification as required by Section 702.4.1.

**302.7 Right of entry.** The code official is authorized to enter the structure or premises at reasonable times to inspect or perform the duties imposed by this code in accordance with Section 702.4.

**302.8 Department records.** The code official shall keep official records of applications received, permits and certificates issued, fees collected, reports of inspections, notices and orders issued, and as required by this code, such records shall be retained in the official records for the period required for retention of public records.

**302.8.1 Approvals and modifications.** A record of approvals and modifications granted shall be maintained by the code official and shall be available for public inspection during business hours in accordance with applicable laws.

**302.8.2 Inspections.** The code official shall keep a record of each inspection made, including notices and orders issued, showing the findings and disposition of each.

**302.8.3 Alternative methods or materials.** The application for modification, alternative methods or materials and the final decision of the code official shall be in writing and shall be officially recorded in the permanent records of the code official.

**302.9 Liability.** The code official, officer or employee charged with the enforcement of this code, while acting for the jurisdiction in good faith and without malice in the discharge of the duties required by this code or other pertinent law or ordinance, shall not thereby be rendered liable personally, and is hereby relieved from all personal liability for any damage accruing to persons or property as a result of any act or by reason of an act or omission in the discharge of official duties. Any suit instituted against any officer or employee because of an act performed by that officer or employee in the lawful discharge of duties and under the provisions of this code shall be defended by the legal representative of the jurisdiction until the final termination of the proceedings.

The code official or any subordinate shall not be liable for costs in any action, suit or proceeding that is instituted in pursuance of the provisions of this code; and any official, officer or employee, acting in good faith and without malice, shall be free from liability for acts performed under any of its provisions or by reason of any act or omission in the performance of official duties in connection therewith.

## SECTION 303
## CERTIFICATE OF OCCUPANCY

**303.1 Use and occupancy.** No building or structure shall be used or occupied until a certificate of occupancy has been provided in accordance with the *International Building Code.*

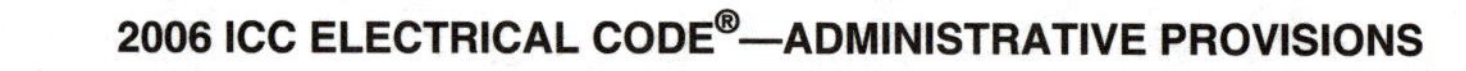

# CHAPTER 4
# PERMITS AND FEES

## SECTION 401 GENERAL

**401.1 Permits required.** Permits required by this code shall be obtained from the code official. Permit fees, if any, shall be paid prior to issuance of the permit. Issued permits shall be kept on the premises designated therein at all times and shall be readily available for inspection by the code official.

**401.2 Types of permits.** An owner, authorized agent or contractor who desires to construct, enlarge, alter, repair, move, demolish or change the occupancy of a building or structure, or to erect, install, enlarge, alter, repair, remove, convert or replace electrical systems or equipment, the installation of which is regulated by this code, or to cause such work to be done, shall first make application to the code official and obtain the required permit for the work.

> **Exception:** Where repair or replacement of electrical systems or equipment must be performed in an emergency situation, the permit application shall be submitted within the next working business day of the department of electrical inspection.

**401.3 Work exempt from permit.** The following work shall be exempt from the requirement for a permit:

1. Listed cord and plug connected temporary decorative lighting.
2. Reinstallation of attachment plug receptacles, but not the outlets therefor.
3. Repair or replacement of branch circuit overcurrent devices of the required capacity in the same location.
4. Temporary wiring for experimental purposes in suitable experimental laboratories.
5. Electrical wiring, devices, appliances, apparatus or equipment operating at less than 25 volts and not capable of supplying more than 50 watts of energy.

Exemption from the permit requirements of this code shall not be deemed to grant authorization for work to be done in violation of the provisions of this code or other laws or ordinances of this jurisdiction.

## SECTION 402 APPLICATION

**402.1 General.** The code official is authorized to receive applications for and issue permits as required by this code.

**402.2 Application.** Application for a permit required by this code shall be made to the code official in such form and detail as prescribed by the code official. Applications for permits shall be accompanied by such plans as prescribed by the code official.

**402.3 Action on application.** The code official shall examine or cause to be examined applications for permits and amendments thereto within a reasonable time after filing. If the application or the construction documents do not conform to the requirements of pertinent laws, the code official shall reject such application in writing, stating the reasons therefor. If the code official is satisfied that the proposed work conforms to the requirements of this code and laws and ordinances applicable thereto, the code official shall issue a permit therefor as soon as practicable.

**402.4 Inspection required.** Before a permit is issued, the code official is authorized to inspect and approve the systems, equipment, buildings, devices, premises and spaces or areas to be used.

**402.5 Time limitation of application.** An application for a permit for any proposed work or operation shall be deemed to have been abandoned 180 days after the date of filing, unless such application has been pursued in good faith or a permit has been issued; except that the code official is authorized to grant one or more extensions of time for additional periods not exceeding 90 days each. The extension shall be requested in writing and justifiable cause demonstrated.

## SECTION 403 CONDITIONS

**403.1 Conditions of a permit.** A permit shall constitute permission to conduct work as set forth in this code in accordance with the provisions of this code. Such permission shall not be construed as authority to violate, cancel or set aside any of the provisions of this code or other applicable regulations or laws of the jurisdiction.

**403.2 Expiration.** Every permit issued shall become invalid unless the work on the site authorized by such permit is commenced within 180 days after its issuance, or if the work authorized on the site by such permit is suspended or abandoned for a period of 180 days after the time the work is commenced. The code official is authorized to grant, in writing, one or more extensions of time, for periods not more than 180 days each. The extension shall be requested in writing and justifiable cause demonstrated.

**403.3 Extensions.** The code official is authorized to grant, in writing, one or more extensions of the time period of a permit for periods of not more than 90 days each. Such extensions shall be requested by the permit holder in writing and justifiable cause demonstrated.

**403.4 Posting the permit.** Issued permits shall be kept on the premises designated therein at all times and shall be readily available for inspection by the code official.

**403.5 Validity.** The issuance or granting of a permit shall not be construed to be a permit for, or an approval of, any violation of any of the provisions of this code or of any other ordinance of the jurisdiction. Permits presuming to give authority to violate or cancel the provisions of this code or other ordinances of the jurisdiction shall not be valid. The issuance of a permit based on construction documents and other data shall not prevent the code official from requiring the correction of errors in the construction documents and other data. The code official is also authorized to prevent occupancy or use of a structure where in violation of this code or of any other ordinances of this jurisdiction.

**403.6 Information on the permit.** The code official shall issue all permits required by this code on an approved form furnished for that purpose. The permit shall contain a general description of the operation or occupancy and its location and any other information required by the code official. Issued permits shall bear the signature of the code official.

**403.7 Suspension or revocation.** The code official is authorized to suspend or revoke a permit issued under the provisions of this code wherever the permit is issued in error, on the basis of incorrect, inaccurate or incomplete information; in violation of any ordinance, regulation or any of the provisions of this code; or if any one of the following conditions exists:

1. The permit is used for a location or establishment other than that for which it was issued.
2. The permit is used for a condition or activity other than that listed in the permit.
3. Conditions and limitations set forth in the permit have been violated.
4. There have been any false statements or misrepresentations as to the material fact in the application for permit or plans submitted or a condition of the permit.
5. The permit is used by a different person or firm than the name for which it was issued.
6. The permittee failed, refused or neglected to comply with orders or notices duly served in accordance with the provisions of this code within the time provided therein.
7. The permit was issued in error or in violation of an ordinance, regulation or this code.

## SECTION 404
## FEES

**404.1 Payment of fees.** A permit shall not be valid until the fees prescribed by law have been paid. Nor shall an amendment to a permit be released until the additional fee, if any, has been paid.

**404.2 Schedule of permit fees.** A fee for each permit shall be paid as required, in accordance with the schedule as established by the applicable governing authority. The fees for electrical work shall be as indicated in the following schedule.

[JURISDICTION TO INSERT APPROPRIATE SCHEDULE]

**404.3 Work commencing before permit issuance.** Any person who commences any work before obtaining the necessary permits shall be subject to an additional fee established by the code official, which shall be in addition to the required permit fees.

**404.4 Related fees.** The payment of the fee for the construction, alteration, removal or demolition for work done in connection with, or concurrently with, the work authorized by a permit shall not relieve the applicant or holder of the permit from the payment of other fees that are prescribed by law.

**404.5 Refunds.** The code official is authorized to establish a refund policy.

# CHAPTER 5
# CONSTRUCTION DOCUMENTS

## SECTION 501
## GENERAL

**501.1 Submittal documents.** Construction documents, special inspection and structural observation programs, and other data shall be submitted in one or more sets with each application for a permit. The construction documents shall be prepared by a registered design professional where required by the statutes of the jurisdiction in which the project is to be constructed. Where special conditions exist, the code official is authorized to require additional construction documents to be prepared by a registered design professional.

**Exception:** The code official is authorized to waive the submission of construction documents and other data not required to be prepared by a registered design professional if it is found that the nature of the work applied for is such that reviewing of construction documents is not necessary to determine compliance with this code.

**501.2 Information on construction documents.** Construction documents shall be drawn to scale upon suitable material. Electronic media documents are permitted to be submitted where approved by the code official. Construction documents shall be of sufficient clarity to indicate the location, nature and extent of the work proposed and show in detail that it will conform to the provisions of this code and relevant laws, ordinances, rules and regulations, as determined by the code official.

**501.2.1 Penetrations.** Construction documents shall indicate where penetrations will be made for electrical systems and shall indicate the materials and methods for maintaining required structural safety, fire-resistance rating and fireblocking.

**501.2.2 Load calculations.** Where an addition or alteration is made to an existing electrical system, an electrical load calculation shall be prepared to determine if the existing electrical service has the capacity to serve the added load.

**501.3 Site plan.** The construction documents submitted with the application for permit shall be accompanied by a site plan showing to scale the size and location of new construction and existing structures on the site, distances from lot lines, the established street grades and the proposed finished grades; and it shall be drawn in accordance with an accurate boundary line survey. In the case of demolition, the site plan shall show construction to be demolished and the location and size of existing structures and construction that are to remain on the site or plot. The code official is permitted to waive or modify the requirement for a site plan where the application for permit is for alteration or repair or where otherwise warranted.

## SECTION 502
## EXAMINATION OF DOCUMENTS

**502.1 General.** The code official shall examine or cause to be examined the accompanying construction documents and shall ascertain by such examinations whether the construction indicated and described is in accordance with the requirements of this code and other pertinent laws or ordinances.

**502.2 Approval of construction documents.** When the code official issues a permit, the construction documents shall be approved, in writing or by stamp, as "Reviewed for Code Compliance." One set of construction documents so reviewed shall be retained by the code official. The other set shall be returned to the applicant, shall be kept at the site of work and shall be open to inspection by the code official or the authorized representative.

**502.2.1 Previous approvals.** This code shall not require changes in the construction documents, construction or installation of electrical systems or equipment for which a lawful permit has been heretofore issued or otherwise lawfully authorized, and the construction of which has been pursued in good faith within 180 days after the effective date of this code and has not been abandoned.

**502.2.2 Phased approval.** The code official is authorized to issue a permit for the installation of part of an electrical system before the construction documents for the electrical system have been submitted, provided that adequate information and detailed statements have been filed complying with pertinent requirements of this code. The holder of such permit shall proceed at the holder's own risk with the building operation and without assurance that a permit for the entire system will be granted.

## SECTION 503
## DESIGN PROFESSIONAL IN RESPONSIBLE CHARGE

**503.1 General.** Where it is required that documents be prepared by a registered design professional, the code official shall require the owner to engage and designate on the permit application a registered design professional who shall act as the registered design professional in responsible charge. If the circumstances require, the owner shall be permitted to designate a substitute registered design professional in responsible charge who shall perform the duties required of the original registered design professional in responsible charge. The code official shall be notified in writing by the owner if the registered design professional in responsible charge is changed or is unable to perform the duties.

The registered design professional in responsible charge shall be responsible for reviewing and coordinating submittal documents prepared by others, including phased and deferred submittal items, for compatibility with the design of the system.

## SECTION 504
## HANDLING SUBMITTALS

**504.1 Deferred submittals.** For the purposes of this section, deferred submittals are defined as those portions of the design that are not submitted at the time of the application and that are to be submitted to the code official within a specified period.

Deferral of any submittal items shall have the prior approval of the code official. The registered design professional in responsible charge shall list the deferred submittals on the construction documents for review by the code official.

Submittal documents for deferred submittal items shall be submitted to the registered design professional in responsible charge, who shall review them and forward them to the code official with a notation indicating that the deferred submittal documents have been reviewed and that they have been found to be in general compliance with the design of the system. The deferred submittal items shall not be installed until their design and submittal documents have been approved by the code official.

**504.2 Amended construction documents.** Work shall be installed in accordance with the reviewed construction documents, and any changes made during construction which are not in compliance with the approved construction documents shall be resubmitted for approval as an amended set of construction documents.

**504.3 Retention of construction documents.** One set of approved construction documents shall be retained by the code official for a period of not less than 180 days from date of completion of the permitted work, or as required by state or local laws.

# CHAPTER 6
# APPROVAL

## SECTION 601
## GENERAL

**601.1 Approved materials and equipment.** All materials, equipment and devices approved by the code official shall be constructed and installed in accordance with such approval.

**601.1.1 Technical assistance.** To determine the acceptability of technologies, processes, products, facilities, materials and uses attending the design, operation or use of a building or premises subject to the inspection of the department, the code official is authorized to require the owner or the person in possession or control of the building or premises to provide, without charge to the jurisdiction, a technical opinion and report. The opinion and report shall be prepared by a qualified engineer, specialist, laboratory or organization acceptable to the code official and shall analyze the properties of the design, operation or use of the building or premises and the facilities and appurtenances situated thereon, to recommend necessary changes. The code official is authorized to require design submittals to be prepared by and bear the stamp of a registered design professional.

**601.2 Modifications.** Wherever there are practical difficulties involved in carrying out the provisions of this code, the code official shall have the authority to grant modifications for individual cases, provided the code official shall first find that special individual reason makes the strict letter of this code impractical and that the modification is in compliance with the intent and purpose of this code, and that such modification does not lessen health, life and fire-safety requirements. The details of action granting modifications shall be recorded and entered in the files of the department of electrical inspection.

**601.3 Alternative materials, methods, equipment and appliances.** The provisions of this code are not intended to prevent the installation of any material or to prohibit any method of construction not specifically prescribed by this code, provided that any such alternative has been approved. An alternative material or method of construction shall be approved where the code official finds that the proposed design is satisfactory and complies with the intent of the provisions of this code, and that the material, method or work offered is, for the purpose intended, at least the equivalent of that prescribed in this code in quality, strength, effectiveness, fire resistance, durability and safety.

**601.4 Material, equipment and appliance reuse.** Materials, equipment, appliances and devices shall not be reused unless such elements have been reconditioned, tested and placed in good and proper working condition and approved.

## SECTION 602
## TESTING

**602.1 Required testing.** Wherever there is insufficient evidence of compliance with the provisions of this code, or evidence that a material or method does not conform to the requirements of this code, or in order to substantiate claims for alternative materials or methods, the code official shall have the authority to require tests as evidence of compliance to be made at no expense to the jurisdiction.

**602.2 Test methods.** Test methods shall be as specified in this code or by other recognized test standards. In the absence of recognized and accepted test methods, the code official shall approve the testing procedures.

**602.3 Testing agency.** All tests shall be performed by an approved agency.

**602.4 Test reports.** Reports of tests shall be retained by the code official for the period required for retention of public records.

## SECTION 603
## ALTERNATIVE ENGINEERED DESIGN

**603.1 General.** The design, documentation, inspection, testing and approval of an alternative engineered design electrical system shall comply with this section.

**603.2 Design criteria.** An alternative engineered design shall conform to the intent of the provisions of this code and shall provide an equivalent level of quality, strength, effectiveness, fire resistance, durability and safety. Materials, equipment or components shall be designed and installed in accordance with the manufacturer's installation instructions.

**603.3 Submittal.** The registered design professional shall indicate on the permit application that the electrical system is an alternative engineered design. The permit and permanent permit records shall indicate that an alternative engineered design was part of the approved installation.

**603.4 Technical data.** The registered design professional shall submit sufficient technical data to substantiate the proposed alternative engineered design and to prove that the performance meets the intent of this code.

**603.5 Construction documents.** The registered design professional shall submit to the code official two complete sets of signed and sealed construction documents for the alternative engineered design. The construction documents shall include floor plans and a diagram of the work.

**603.6 Design approval.** Where the code official determines that the alternative engineered design conforms to the intent of this code, the electrical system shall be approved. If the alternative engineered design is not approved, the code official shall notify the registered design professional in writing, stating the reasons therefor.

**603.7 Inspection and testing.** The alternative engineered design shall be tested and inspected in accordance with the requirements of this code.

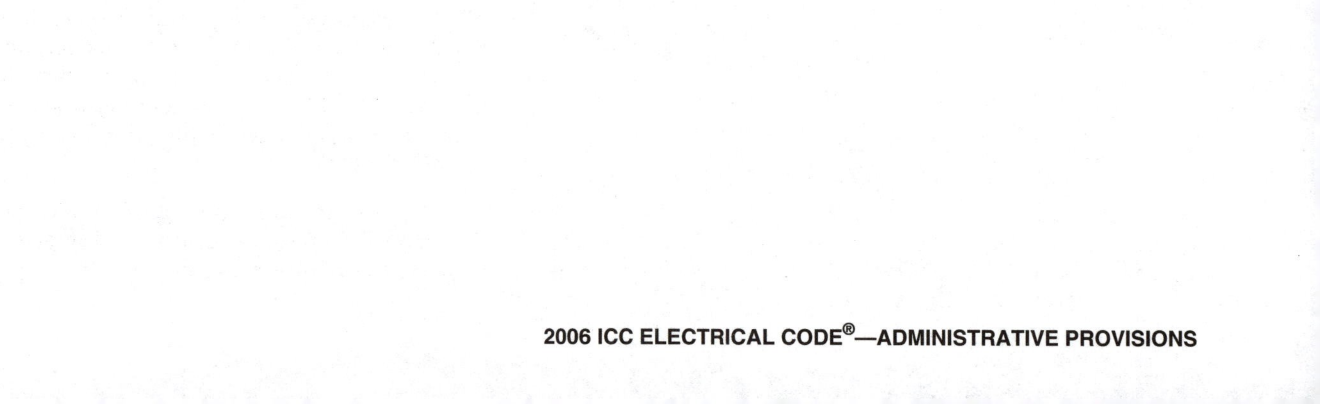

# CHAPTER 7
# INSPECTIONS AND TESTING

## SECTION 701
## GENERAL

**701.1 General.** The code official is authorized to conduct inspections that are deemed necessary to determine the extent of compliance with the provisions of this code and to approve reports of inspection by approved agencies or individuals. All reports of such inspections shall be prepared and submitted in writing for review and approval. Inspection reports shall be certified by a responsible officer of such approved agency or by the responsible individual. The code official is authorized to engage such expert opinion as deemed necessary to report upon unusual, detailed or complex technical issues subject to the approval of the governing body.

## SECTION 702
## REQUIRED INSPECTIONS

**702.1 General.** The code official, upon notification, shall make the inspections set forth in this section.

**702.1.1 Evaluation and follow-up inspection services.** Prior to the approval of a prefabricated construction assembly having concealed electrical work and the issuance of an electrical permit, the code official shall require the submittal of an evaluation report on each prefabricated construction assembly, indicating the complete details of the electrical system, including a description of the system and its components, the basis upon which the system is being evaluated, test results and similar information, and other data as necessary for the code official to determine conformance to this code.

**702.1.1.1 Evaluation service.** The code official shall designate the evaluation service of an approved agency as the evaluation agency, and review such agency's evaluation report for adequacy and conformance to this code.

**702.1.1.2 Follow-up inspection.** Except where ready access is provided to electrical systems, service equipment and accessories for complete inspection at the site without disassembly or dismantling, the code official shall conduct the in-plant inspections as frequently as necessary to ensure conformance to the approved evaluation report or shall designate an independent, approved inspection agency to conduct such inspections. The inspection agency shall furnish the code official with the follow-up inspection manual and a report of inspections upon request, and the electrical system shall have an identifying label permanently affixed to the system indicating that factory inspections have been performed.

**702.1.1.3 Test and inspection records.** Required test and inspection records shall be available to the code official at all times during the fabrication of the electrical system and the erection of the building; or such records as the code official designates shall be filed.

**702.1.2 Concealed work.** Work shall remain accessible and exposed for inspection purposes until approved. It shall be the duty of the permit applicant to cause the work to remain accessible and exposed for inspection purposes. Wherever any installation subject to inspection prior to use is covered or concealed without having first been inspected, the code official shall have the authority to require that such work be exposed for inspection. Neither the code official nor the jurisdiction shall be liable for expense entailed in the removal or replacement of any material required to allow inspection.

**702.1.3 Underground.** Underground inspection shall be made after trenches or ditches are excavated and bedded, piping and conductors installed, and before backfill is put in place. Where excavated soil contains rocks, broken concrete, frozen chunks and other rubble that would damage or break the raceway, cable or conductors, or where corrosive action will occur, protection shall be provided in the form of granular or selected material, approved running boards, sleeves or other means.

**702.1.4 Rough-in.** Rough-in inspection shall be made after the roof, framing, fireblocking and bracing are in place and all wiring and other components to be concealed are complete, and prior to the installation of wall or ceiling membranes.

**702.1.5 Other inspections.** In addition to the inspections specified above, the code official is authorized to make or require other inspections of any construction work to ascertain compliance with the provisions of this code and other laws, which are enforced by the department of electrical inspection.

**702.1.6 Final inspection.** The final inspection shall be made after all work required by the permit is completed.

**702.1.7 Inspection record card.** Work requiring a permit shall not be commenced until the permit holder or an agent of the permit holder shall have posted or otherwise made available an inspection record card such as to allow the code official to make conveniently the required entries thereon regarding inspection of the work. This card shall be maintained by the permit holder until final approval has been granted by the code official.

**702.1.8 Approval required.** Work shall not be performed beyond the point indicated in each successive inspection and test without first obtaining the approval of the code official. The code official, upon notification, shall make the requested inspections and tests and shall either indicate the portion of the construction that is satisfactory as completed, or shall notify the permit holder or an agent of the permit holder wherein the same fails to comply with this code. Any portions that do not comply shall be corrected and such portion shall not be covered or concealed until authorized by the code official.

**702.2 Validity.** Approval as a result of an inspection shall not be construed to be an approval of a violation of the provisions of this code or of other ordinances of the jurisdiction. Inspections presuming to give authority to violate or cancel the provisions of this code or of other ordinances of the jurisdiction shall not be valid.

**702.3 Preliminary inspection.** Before issuing a permit, the code official is authorized to examine or cause to be examined buildings, structures and sites for which an application has been filed. The code official shall be notified when the installation is ready for inspection and is authorized to conduct the inspection within a reasonable period of time.

**702.4 Entry.** The code official is authorized to enter and examine any building, structure, marine vessel, vehicle or premises in accordance with Section 702.4.3 for the purpose of enforcing this code.

**702.4.1 Identification.** The code official shall carry proper identification issued by the governing authority where inspecting structures, premises or facilities in the performance of duties under this code and shall be identified by proper credentials issued by this governing authority.

**702.4.2 Impersonation prohibited.** A person shall not impersonate the code official through the use of a uniform, identification card, badge or any other means.

**702.4.3 Right of entry.** Where it is necessary to make an inspection to enforce the provisions of this code, or where the code official has reasonable cause to believe that there exists in a structure or upon any premises a condition that is contrary to or in violation of this code, which makes the structure or premises unsafe, dangerous or hazardous, the code official is authorized to enter the structure or premises at reasonable times to inspect or to perform the duties imposed by this code, provided that if such structure or premises be occupied, that credentials be presented to the occupant and entry requested. If such structure or premises is unoccupied, the code official is authorized to first make a reasonable effort to locate the owner or other person having charge or control of the structure or premises and request entry. If entry is refused, the code official shall have recourse to the remedies provided by law to secure entry.

**702.5 Inspection agencies.** The code official is authorized to accept reports of approved inspection agencies, provided such agencies satisfy the requirements as to qualifications and reliability.

**702.6 Inspection requests.** It shall be the duty of the person doing the work authorized by a permit to notify the code official that such work is ready for inspection. It shall be the duty of the person requesting any inspections required by this code to provide access to and means for inspection of such work.

**702.7 Assistance from other agencies.** The assistance and cooperation of police, building, fire and health department officials and all other officials shall be available as required in the performance of duties.

**702.8 Contractors' responsibilities.** It shall be the responsibility of every contractor who enters into contracts for the installation or repair of electrical systems for which a permit is required to comply with adopted state and local rules and regulations concerning licensing.

## SECTION 703 TESTING

**703.1 General.** Electrical work shall be tested as required in this code. Tests shall be performed by the permit holder and observed by the code official.

**703.2 Apparatus, material and labor for tests.** Apparatus, material and labor required for testing an electrical system or part thereof shall be furnished by the permit holder.

**703.3 Reinspection and testing.** Where any work or installation does not pass an initial test or inspection, the necessary corrections shall be made so as to achieve compliance with this code. The work or installation shall then be resubmitted to the code official for inspection and testing.

# CHAPTER 8
# SERVICE UTILITIES

## SECTION 801
## GENERAL

**801.1 Connection of service utilities.** No person shall make connections from a utility, source of energy, fuel or power to any building or system that is regulated by this code for which a permit is required, until released by the code official.

**801.2 Temporary connection.** The code official shall have the authority to authorize the temporary connection of the building or system to the utility source of energy, fuel or power.

**801.3 Authority to disconnect service utilities.** The code official shall have the authority to authorize disconnection of utility services or energy sources to the building, structure or system regulated by this code in case of an emergency where it is necessary to eliminate an immediate hazard to life or property. The code official shall notify the serving utility and, wherever possible, the owner and occupant of the building, structure or service system of the decision to disconnect prior to taking such action. If not notified prior to disconnecting, the owner or occupant of the building, structure or service system shall be notified in writing as soon as practical thereafter.

**801.3.1 Connection after order to disconnect.** A person shall not make utility service or energy source connections to systems regulated by this code, which have been disconnected or ordered to be disconnected by the code official, or the use of which has been ordered to be discontinued by the code official until the code official authorizes the reconnection and use of such systems.

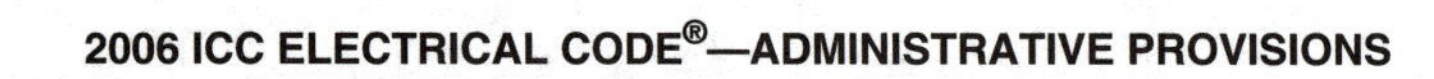

# CHAPTER 9
# UNSAFE SYSTEMS AND EQUIPMENT

## SECTION 901
## CONDITIONS

**901.1 Unsafe electrical systems.** An electrical system that is unsafe, constitutes a fire or health hazard, or is otherwise dangerous to human life, as regulated by this code, is hereby declared as an unsafe electrical system. Use of an electrical system regulated by this code constituting a hazard to health, safety or welfare by reason of inadequate maintenance, dilapidation, fire hazard, disaster, damage or abandonment is hereby declared an unsafe use. Such unsafe equipment and appliances are hereby declared to be a public nuisance and shall be abated by repair, rehabilitation, demolition or removal.

**901.2 Authority to condemn electrical systems.** Wherever the code official determines that any electrical system, or portion thereof, regulated by this code has become hazardous to life, health or property, the code official shall order in writing that such electrical systems either be removed or restored to a safe condition. A time limit for compliance with such order shall be specified in the written notice. A person shall not use or maintain a defective electrical system or equipment after receiving such notice.

Where such electrical system is to be disconnected, written notice as prescribed in this code shall be given. In cases of immediate danger to life or property, such disconnection shall be made immediately without such notice.

**901.3 Dangerous conditions.** Wherever the code official shall find in any structure or upon any premises dangerous or hazardous conditions or materials, the code official is authorized to order such dangerous conditions or materials to be removed or remedied in accordance with the provisions of this code.

**901.4 Record.** The code official shall cause a report to be filed on an unsafe condition. The report shall state the occupancy of the structure and the nature of the unsafe condition.

**901.5 Notice.** If an unsafe condition is found, the code official shall serve on the owner, agent or person in control of the structure, a written notice that describes the condition deemed unsafe and specifies the required repairs or improvements to be made to abate the unsafe condition, or that requires the unsafe condition to be removed within a stipulated time. Such notice shall require the person thus notified to declare immediately to the code official acceptance or rejection of the terms of the order.

**901.6 Method of service.** Such notice shall be deemed properly served if a copy thereof is: (a) delivered to the owner personally; or (b) sent by certified or registered mail addressed to the owner at the last known address with the return receipt requested. If the certified or registered letter is returned showing that the letter was not delivered, a copy thereof shall be posted in a conspicuous place in or about the structure affected by such notice. Service of such notice in the foregoing manner upon the owner's agent or upon the person responsible for the structure shall constitute service of notice upon the owner.

# CHAPTER 10

# VIOLATIONS

## SECTION 1001 UNLAWFUL ACTS

**1001.1 General.** It shall be unlawful for any person, firm or corporation to erect, construct, alter, extend, repair, move, remove, demolish or occupy any system or equipment regulated by this code, or cause same to be done, in conflict with or in violation of any of the provisions of this code.

## SECTION 1002 NOTICE OF VIOLATION

**1002.1 Issuance.** Where the code official finds any building, premises, vehicle, system or equipment that is in violation of this code, the code official is authorized to issue corrective orders.

**1002.2 Notice.** Wherever the code official determines violations of this code or observes an apparent or actual violation of a provision of this code or other codes or ordinances under the code official's jurisdiction, the code official is authorized to prepare a written notice of violation describing the conditions deemed unsafe and, where compliance is not immediate, specifying a time for reinspection. Such order shall direct the discontinuance of the illegal action or condition and the abatement of the violation.

**1002.3 Service.** Any order or notice issued pursuant to this code shall be served upon the owner, operator, occupant or other person responsible for the condition or violation, either by personal service, mail or by delivering the same to, and leaving it with, some person of responsibility upon the premises. For unattended or abandoned locations, a copy of such order or notice shall be posted on the premises in a conspicuous place at or near the entrance to such premises, and the order or notice shall be mailed by certified mail with return receipt requested or a certificate of mailing, to the last known address of the owner, occupant or both.

**1002.4 Compliance with orders and notices.** Orders and notices issued or served as provided by this code shall be complied with by the owner, operator, occupant or other person responsible for the condition or violation to which the order or notice pertains.

**1002.5 Failure to correct violations.** If the notice of violation is not complied with, the code official is authorized to request the legal counsel of the jurisdiction to institute the appropriate legal proceedings to restrain, correct or abate such violation or to require removal or termination of the unlawful occupancy of the structure in violation of the provisions of this code or of any order or direction made pursuant thereto.

**1002.6 Failure to comply.** Failure to comply with an abatement notice or other corrective notice issued by the code official shall result in each day that such violation continues being regarded as a new and separate offense.

**1002.7 Unauthorized tampering.** Signs, tags or seals posted or affixed by the code official shall not be mutilated, destroyed or tampered with or removed without authorization from the code official.

## SECTION 1003 PENALTIES

**1003.1 Penalties.** Any person who fails to comply with the provisions of this code or who fails to carry out an order made pursuant of this code or violates any condition attached to a permit, approval or certificate shall be subject to the penalties established by this jurisdiction.

**1003.2 Abatement of violation.** The imposition of the penalties herein described shall not prevent the legal officer of the jurisdiction from instituting appropriate action to prevent unlawful construction or to restrain, correct or abate a violation; or to prevent illegal occupancy of a structure or premises; or to stop an illegal act, conduct of business or occupancy of a structure on or about any premises.

## SECTION 1004 STOP WORK ORDER

**1004.1 Issuance.** Upon notice from the code official that any electrical work is being done contrary to the provisions of this code or in a dangerous or unsafe manner, such work shall immediately cease. Such notice shall be in writing and shall be given to the owner of the property, or to the owner's agent, or to the person doing the work. The notice shall state the conditions under which work is authorized to resume.

**1004.2 Emergencies.** Where an emergency exists, the code official shall not be required to give a written notice prior to stopping the work.

**1004.3 Unlawful continuance.** Any person who shall continue any work in or about the structure after having been served with a stop work order, except such work as that person is directed to perform to remove a violation or unsafe condition, shall be subject to penalties as prescribed by law.

# CHAPTER 11
# MEANS OF APPEAL

## SECTION 1101
## GENERAL

**1101.1 Board of appeals established.** In order to hear and decide appeals of orders, decisions or determinations made by the code official relative to the application and interpretation of this code, there shall be and is hereby created a board of appeals. The board of appeals shall be appointed by the governing body and shall hold office at its pleasure. The board shall adopt rules of procedure for conducting its business, and shall render all decisions and findings in writing to the appellant with a duplicate copy to the code official.

**1101.2 Limitations on authority.** An application for appeal shall be based on a claim that the true intent of this code or the rules legally adopted thereunder have been incorrectly interpreted, the provisions of this code do not fully apply, or an at least equivalent method of protection or safety is proposed. The board shall have no authority to waive the requirements of this code.

## SECTION 1102
## MEMBERSHIP

**1102.1 Membership of board.** The board of appeals shall consist of five members appointed by the chief appointing authority as follows: one for five years; one for four years; one for three years; one for two years and one for one year. Thereafter, each new member shall serve for five years or until a successor has been appointed.

**1102.2 Qualifications.** The board of appeals shall consist of five individuals, one from each of the following professions or disciplines:

1. Registered design professional who is a registered architect; or a builder or superintendent of building construction with at least 10 years' experience, five of which shall have been in responsible charge of work.
2. Registered design professional with structural engineering or architectural experience.
3. Registered design professional with mechanical, plumbing or fuel-gas engineering experience; or a mechanical, plumbing or fuel-gas contractor with at least 10 years' experience, five of which shall have been in responsible charge of work.
4. Registered design professional with electrical engineering experience; or an electrical contractor with at least 10 years' experience, five of which shall have been in responsible charge of work.
5. Registered design professional with fire protection engineering experience; or a fire protection contractor with at least 10 years' experience, five of which shall have been in responsible charge of work.
6. The code official shall be an ex officio member of said board, but shall have no vote on any matter before the board.

**1102.3 Alternate members.** The chief appointing authority shall appoint two alternate members who shall be called on by the board chairman to hear appeals during the absence or disqualification of a member. Alternate members shall possess the qualifications required for board membership and shall be appointed for five years, or until a successor has been appointed.

**1102.4 Chairman.** The board shall annually select one of its members to serve as chairman.

**1102.5 Disqualification of members.** A member shall not hear an appeal in which that member has a personal, professional or financial interest.

**1102.6 Secretary.** The chief appointing authority shall designate a qualified clerk to serve as secretary to the board. The secretary shall file a detailed record of all proceedings in the office of the chief appointing authority.

**1102.7 Compensation of members.** Compensation of members shall be determined by law.

## SECTION 1103
## PROCEDURES

**1103.1 Application for appeal.** A person shall have the right to appeal a decision of the code official to the board of appeals. An application for appeal shall be based on a claim that the true intent of this code or the rules legally adopted thereunder, have been incorrectly interpreted, the provisions of this code do not fully apply, or an equally good or better form of construction is proposed. The application shall be filed on a form obtained from the code official within 20 days after the notice was served.

**1103.2 Notice of meeting.** The board shall meet upon notice from the chairman, within 10 days of the filing of an appeal, or at stated periodic meetings.

**1103.3 Open hearing.** All hearings before the board shall be open to the public. The appellant, the appellant's representative, the code official and any person whose interests are affected shall be given an opportunity to be heard.

**1103.4 Rules of procedure.** The board shall adopt and make available to the public through the secretary rules of procedure under which a hearing will be conducted. The procedures shall not require compliance with strict rules of evidence, but shall mandate that only relevant information be received.

**1103.5 Postponed hearing.** Where five members are not present to hear an appeal, either the appellant or the appellant's representative shall have the right to request a postponement of the hearing.

**1103.6 Decisions.** The board shall modify or reverse the decision of the code official by a concurring vote of three members.

**1103.6.1 Resolution.** The decision of the board shall be by resolution. Certified copies shall be furnished to the appellant and to the code official.

**1103.6.2 Administration.** The code official shall take immediate action in accordance with the decision of the board.

**1103.7 Court review.** Any person, whether or not a previous party of the appeal, shall have the right to apply to the appropriate court for a writ of certiorari to correct errors of law. Application for review shall be made in the manner and time required by law following the filing of the decision in the office of the chief appointing authority.

# CHAPTER 12
# ELECTRICAL PROVISIONS

## SECTION 1201
## GENERAL

**1201.1 Scope.** This chapter governs the design and construction of electrical systems and equipment.

**1201.1.1 Adoption.** Electrical systems and equipment shall be designed and constructed in accordance with the *International Residential Code* or NFPA 70 as applicable, except as otherwise provided in this code.

**[F] 1201.2 Abatement of electrical hazards.** All identified electrical hazards shall be abated. All identified hazardous electrical conditions in permanent wiring shall be brought to the attention of the code official responsible for enforcement of this code. Electrical wiring, devices, appliances and other equipment which is modified or damaged and constitutes an electrical shock or fire hazard shall not be used.

**[F] 1201.3 Appliance and fixture listing.** Electrical appliances and fixtures shall be tested and listed in published reports of inspected electrical equipment by an approved agency and installed in accordance with all instructions included as part of such listing.

## SECTION 1202
## PROVISIONS

**1202.1 General.** The provisions of this section shall apply to the design, construction, installation, use and maintenance of electrical systems and equipment. Where differences occur between provisions of this code and referenced codes or standards, the provisions of this code shall apply.

**1202.2 Nonmetallic-sheathed cable.** The use of Type NM, NMC and NMS (nonmetallic sheathed) cable wiring methods shall not be limited based on height, number of stories or construction type of the building or structure.

**1202.3 Cutting, notching and boring.** The cutting, notching and boring of wood and steel framing members, structural members and engineered wood products shall be in accordance with the *International Building Code*.

**1202.4 Penetrations.** Penetrations of walls, floors, ceilings and assemblies required to have a fire-resistance rating, shall be protected in accordance with the *International Building Code*. Where cables, conductors and raceways penetrate fireblocking or draftstopping, such penetrations shall be protected by filling the annular space with an approved fireblocking material.

**1202.5 Smoke detector circuits.** Smoke detectors required by the *International Building Code* and installed within dwelling units shall not be connected as the only load on a branch circuit. Such detectors shall be supplied by branch circuits having lighting loads consisting of lighting outlets in habitable spaces.

**[M] 1202.6 Appliance access.** Where appliances requiring access are installed in attics or underfloor spaces, a luminaire controlled by a switch located at the required passageway opening to such space and a receptacle outlet shall be provided at or near the appliance location.

**[FG] 1202.7 Prohibited grounding electrode.** Fuel gas piping shall not be used as a grounding electrode.

**[F] 1202.8 Emergency and standby power.** Emergency and standby power systems required by the *International Building Code* or *International Fire Code* shall be installed in accordance with the *International Building Code,* the *International Fire Code,* NFPA 110, NFPA 111 and this code.

**[F] 1202.9 Smoke control systems.** Smoke control systems required by the *International Building Code* or *International Fire Code* shall be supplied with two sources of power. Primary power shall be the normal building power systems. Secondary power shall be from an approved standby source complying with this code. The standby power source and its transfer switches shall be in a separate room from the normal power transformers and switch gear, and shall be enclosed in a room constructed of not less than 1-hour fire-resistance-rated fire barriers, ventilated directly to and from the exterior. Power distribution from the two sources shall be by independent routes. Transfer to full standby power shall be automatic and within 60 seconds of failure of the primary power.

**[F] 1202.9.1 Power sources and power surges.** Elements of the smoke management system relying on volatile memories or the like shall be supplied with uninterruptable power sources of sufficient duration to span 15-minute primary power interruption. Elements of the smoke management system susceptible to power surges shall be suitably protected by conditioners, suppressors or other approved means.

**[F] 1202.9.2 Wiring.** In addition to meeting the requirements of this code, all signal and control wiring for smoke control systems, regardless of voltage, shall be fully enclosed within continuous raceways.

**[M] 1202.10 Wiring in plenums.** Combustible electrical or electronic wiring methods and materials, optical fiber cable, and optical fiber raceway exposed within plenums regulated by Section 602 of the *International Mechanical Code* shall have a peak optical density not greater than 0.50, an average optical density not greater than 0.15, and a flame spread not greater than 5 feet (1524 mm) when tested in accordance with NFPA 262. Only type OFNP (plenum-rated nonconducive optical fiber cable) shall be installed in plenum-rated optical fiber raceways. Wiring, cable and raceways addressed in this section shall be listed and labeled as plenum rated and shall be installed in accordance with this code.

**[M] 1202.10.1 Combustible electrical equipment.** Combustible electrical equipment exposed within plenums regulated by Section 602 of the *International Mechanical Code* shall have a peak rate of heat release not greater than 100 kilowatts (kW), a peak optical density not greater than 0.50, and an average optical density not greater than 0.15 when tested in accordance with UL 2043. Combustible electrical equipment shall be listed and labeled.

**[M] 1202.11 Engine and gas turbine-powered equipment and appliances.** Permanently installed equipment and appliances powered by internal combustion engines and turbines shall be installed in accordance with the manufacturer's installation instructions, the *International Mechanical Code, International Fuel Gas Code* and NFPA 37.

**[F] 1202.12 Stationary fuel cell power systems.** Stationary fuel cell power systems having a power output not exceeding 10 MW shall be tested in accordance with ANSI CSA America FC1 and shall be installed in accordance with the manufacturer's installation instructions and NFPA 853.

**[M] 1202.13 Boiler control requirements.** The power supply to the electrical control system for boilers shall be from a two-wire branch circuit that has a grounded conductor or from an isolation transformer with a two-wire secondary. Where an isolation transformer is provided, one conductor of the secondary winding shall be grounded. Control voltage shall not exceed 150 volts nominal, line to line. Control and limit devices shall interrupt the ungrounded side of the circuit. A means of manually disconnecting the control circuit shall be provided, and controls shall be arranged so that when deenergized, the burner shall be inoperative. Such disconnecting means shall be capable of being locked in the off position and shall be provided with ready access.

**[F] 1202.14 Equipment and door labeling.** Doors into electrical control panel rooms shall be marked with a plainly visible and legible sign stating ELECTRICAL ROOM or similar approved wording. The disconnecting means for each service, feeder or branch circuit originating on a switchboard or panelboard shall be legibly and durably marked to indicate its purpose unless such purpose is clearly evident.

**[F] 1202.15 Smoke alarm power source.** In new construction, required smoke alarms shall receive their primary power from the building wiring where such wiring is served from a commercial source and shall be equipped with a battery backup. Smoke alarms shall emit a signal when the batteries are low. Wiring shall be permanent and without a disconnecting switch other than as required for overcurrent protection.

**Exception:** Smoke alarms are not required to be equipped with battery backup in Group R-1 where they are connected to an emergency electrical system.

**[F] 1202.16 Smoke alarm interconnection.** Where more than one smoke alarm is required to be installed within an individual dwelling unit or sleeping unit in Group R-2, R-3 or R-4, or within an individual sleeping unit in Group R-1, the smoke alarms shall be interconnected in such a manner that the activation of one alarm will activate all of the alarms in the individual unit. The alarm shall be clearly audible in all bedrooms over background noise levels with all intervening doors closed.

## SECTION 1203
## EXISTING ELECTRICAL FACILITIES

**[PM] 1203.1 Existing buildings.** This section shall apply to buildings and structures that are within the scope of the *International Property Maintenance Code.* Every occupied building shall be provided with an electrical system in compliance with the requirements of Sections 1203.1.1 through 1203.1.5.

**[PM] 1203.1.1 Service.** The size and usage of appliances and equipment shall serve as a basis for determining the need for additional facilities in accordance with this code. Dwelling units shall be served by a three-wire, 120/240 volt, single-phase electrical service having a rating of not less than 60 amperes.

**[PM] 1203.1.2 Electrical system hazards.** Where it is found that the electrical system in a structure constitutes a hazard to the occupants or the structure by reason of inadequate service, improper fusing, insufficient receptacle and lighting outlets, improper wiring or installation, deterioration or damage, or for similar reasons, the code official shall require the defects to be corrected to eliminate the hazard.

**[PM] 1203.1.3 Installation.** All electrical equipment, wiring and appliances shall be properly installed and maintained in a safe and approved manner.

**[PM] 1203.1.4 Receptacles.** Every habitable space in a dwelling shall be provided with at least two separate and remote receptacle outlets. Every laundry area shall be provided with at least one grounding-type receptacle outlet or a receptacle outlet with ground fault circuit interrupter protection. Every bathroom shall contain at least one receptacle outlet. Any new bathroom receptacle outlet shall have ground fault circuit interrupter protection.

**[PM] 1203.1.5 Luminaires.** Every public hall, interior stairway, toilet room, kitchen, bathroom, laundry room, boiler room and furnace room shall be provided with at least one electric luminaire.

# CHAPTER 13

# REFERENCED STANDARDS

This chapter lists the standards that are referenced in various sections of this document. The standards are listed herein by the promulgating agency of the standard, the standard identification, the effective date and title, and the section or sections of this document that reference the standard. The application of the referenced standards shall be as specified in Section 102.6.

## ICC

International Code Council
Suite 600
5203 Leesburg Pike
Falls Church, VA 22041

| Standard reference number | Title | Referenced in code section number |
|---|---|---|
| IBC—06 | International Building Code® | 201.3, 303.1, 1202.3, 1202.4, 1202.5, 1202.8, 1202.9 |
| IECC—06 | International Energy Conservation Code® | 201.3 |
| IFC—06 | International Fire Code® | 201.3, 1202.8, 1202.9 |
| IFGC—06 | International Fuel Gas Code® | 201.3, 1202.11 |
| IMC—06 | International Mechanical Code® | 201.3, 1202.10, 1202.10.1, 1202.11 |
| IPC—06 | International Plumbing Code® | 201.3 |
| IPMC—06 | International Property Maintenance Code® | 201.3, 1203.1 |
| IPSDC—06 | International Private Sewage Disposal Code® | 201.3 |
| IRC—06 | International Residential Code® | 201.3, 1201.1.1 |
| IZC—06 | International Zoning Code® | 201.3 |

## NFPA

National Fire Protection Association
1 Batterymarch Park
Quincy, MA 02269

| Standard reference number | Title | Referenced in code section number |
|---|---|---|
| 37—02 | Installation and Use of Stationary Combustion Engines and Gas Turbines | 1202.11 |
| 70—05 | National Electrical Code | 201.3, 1201.1.1 |
| 110—99 | Emergency and Standby Power Systems | 1202.8 |
| 111—01 | Stored Electrical Energy Emergency and Standby Power Systems | 1202.8 |
| 262—99 | Standard Method of Test for Flame Travel and Smoke of Wires and Cables for Use in Air-Handling Spaces | 1202.10 |
| 853—03 | Installation of Stationary Fuel Cell Power Systems | 1202.12 |

## UL

Underwriters Laboratories
333 Pfingsten Road
Northbrook, IL 60062–2096

| Standard reference number | Title | Referenced in code section number |
|---|---|---|
| 2043—96 | Fire Test for Heat and Visible Smoke Release for Discrete Products and their Accessories Installed in Air-Handling Spaces—with Revisions Through February 1998 | 1202.10.1 |

# INDEX


## A

**ADMINISTRATION**
- Additions, alterations and repairs . . . . . . . . . . 102.1.3
- Change in occupancy . . . . . . . . . . . . . . . . . . 102.1.4
- Existing installations . . . . . . . . . . . . . . . . . . 102.1.1
- Maintenance . . . . . . . . . . . . . . . . . . . . . . . . 102.1.2
- Means of Appeal . . . . . . . . . . . . . . . . . . . . . . . 1101
- Referenced codes and standards . . . . . . . . . . . 102.6

**ALTERNATE METHODS OR MATERIALS** . . . . . . . . . . . . . . . . . . . . . . . 302.8.3, 601.3, 603.1

**APPROVAL AND MODIFICATIONS** . . . . . . . . . 302.8.1

**APPROVED** . . . . . . . . . . . . . . . . . . . . . . . . . . . . 202

**APPROVED AGENCY** . . . . . . . . . . . . . . . . . . . . . 202

## C

**CODE OFFICIAL**
- Appointment . . . . . . . . . . . . . . . . . . . . . . . . . 301.2
- Definition . . . . . . . . . . . . . . . . . . . . . . . . . . . . 202
- Authority . . . . . . . . . . . . . . . . . . . . . . . . . . . 302.1
- Duties . . . . . . . . . . . . . . . . . . . . . . . . . . . . . 302.2
- Identification . . . . . . . . . . . . . . . . . . . . . . . . 302.6
- Inspections . . . . . . . . . . . . . . . . . . . . . . . . . 302.5
- Liability . . . . . . . . . . . . . . . . . . . . . . . . . . . . 302.9

**CERTIFICATE OF OCCUPANCY**
- Change in use . . . . . . . . . . . . . . . . . . . . . . . 102.1.4
- Use and occupancy . . . . . . . . . . . . . . . . . . . . 303.1

**CONSTRUCTION DOCUMENTS** . . . . . . . . . . . . . 501.1

## D

**DESIGN PROFESSIONAL** . . . . . . . . . . . . . . . . . 503.1

## E

**ELECTRICAL SYSTEMS**
- Authority to condemn . . . . . . . . . . . . . . . . . . . 901.2
- Dangerous conditions . . . . . . . . . . . . . . . . . . . 901.3
- Existing systems . . . . . . . . . . . . . . . . . . . . . . 1203.1
- General . . . . . . . . . . . . . . . . . . . . . . . . . . . . 1202.1
- Method of service . . . . . . . . . . . . . . . . . . . . . 901.6
- Smoke detector circuits . . . . . . . . . . . . . . . . . 1202.5
- Unsafe . . . . . . . . . . . . . . . . . . . . . . . . . . . . . 901.1

**EQUIPMENT AND DOOR LABELING** . . . . . . . 1202.14

**EXISTING ELECTRICAL FACILITIES**
- Installation . . . . . . . . . . . . . . . . . . . . . . . . 1203.1.3
- Receptacles . . . . . . . . . . . . . . . . . . . . . . . . 1203.1.4
- Luminaires . . . . . . . . . . . . . . . . . . . . . . . . . 1203.1.5

## I

**INSPECTION AND TESTING**
- Concealed work . . . . . . . . . . . . . . . . . . . . . . 702.1.2
- Final . . . . . . . . . . . . . . . . . . . . . . . . . . . . . 702.1.6
- Required . . . . . . . . . . . . . . . . . . . . . . . . . . . . 702.1
- Rough-in . . . . . . . . . . . . . . . . . . . . . . . . . . . 702.1.4
- Underground . . . . . . . . . . . . . . . . . . . . . . . . 702.1.3

## L

**LISTED AND LISTING**
- Appliance and fixture . . . . . . . . . . . . . . . . . . . 1201.3
- Definition . . . . . . . . . . . . . . . . . . . . . . . . . . . . 202

## O

**OCCUPANCY** . . . . . . . . . . . . . . . . . . . . . . . . . . . 202

## P

**PERMIT**
- Application for . . . . . . . . . . . . . . . . . . . 302.3, 402.2
- Conditions of . . . . . . . . . . . . . . . . . . . . . . . . 403.1
- Exempt from . . . . . . . . . . . . . . . . . . . . . . . . . 401.3
- Fees . . . . . . . . . . . . . . . . . . . . . . . . . . . . . . . 404

## S

**SCOPE** . . . . . . . . . . . . . . . . . . . . . . . . . . . . . 101.3

**SERVICE UTILITIES**
- Authority to disconnect . . . . . . . . . . . . . . . . . 801.3
- Connection . . . . . . . . . . . . . . . . . . . . . . . . . . 801.1
- Temporary connection . . . . . . . . . . . . . . . . . . 801.2

## V

**VIOLATIONS**
- Failure to correct . . . . . . . . . . . . . . . . . . . . . 1002.5
- Notice of . . . . . . . . . . . . . . . . . . . . . . . . . . . 1002.2
- Penalties . . . . . . . . . . . . . . . . . . . . . . . . . . . 1003.1
- Stop work order . . . . . . . . . . . . . . . . . . . . . . 1004.1

## W

**WIRING**
- NM Cable . . . . . . . . . . . . . . . . . . . . . . . . . . . 1202.2
- Plenums . . . . . . . . . . . . . . . . . . . . . . . . . . . 1202.10
- Smoke control systems . . . . . . . . . . . . . . . . . 1202.9

*People Helping People Build a Safer World™*

# Membership Application

*This form may be photocopied*

## Member Categories and Dues*

Special membership structures are also available for Educational and Federal Agencies.
For more information, please visit www.iccsafe.org/membership or call 1-888-ICC-SAFE (422-7233), ext.33804.

## GOVERNMENTAL MEMBER**

Government/Municipality (including agencies, departments or units) engaged in administration, formulation or enforcement of laws, regulations or ordinances relating to public health, safety and welfare. Annual member dues (by population) are shown below. Please verify the current ICC membership status of your employer prior to applying.

☐ Up to 50,000..........$100 ☐ 50,001-150,000........ $180 ☐ 150,001+.......... $280

**A Governmental Member may designate four to 12 voting representatives (based on population) who are employees or officials of that governmental member and are actively engaged on a full- or part-time basis in the administration, formulation or enforcement of laws, regulations or ordinances relating to public health, safety and welfare. Number of representatives is based on population. All fees for representatives have been included in the annual member dues payment. Visit www.iccsafe.org/membership for more details.

☐ **CORPORATE MEMBERS ($300)** An association, society, testing laboratory, manufacturer, company or corporation.

**INDIVIDUAL MEMBERS**

☐ **PROFESSIONAL ($150)** A design professional duly licensed or registered by any state or other recognized governmental agency.

☐ **COOPERATING ($150)** An individual who is interested in International Code Council purposes and objectives and would like to take advantage of membership benefits.

☐ **CERTIFIED ($75)** An individual who holds a current Legacy or International Code Council certification.

☐ **ASSOCIATE ($35)** An employee of a Governmental Member who may or may not be a Governmental Member Voting Representative.

☐ **STUDENT ($25)** An individual who is enrolled in classes or a course of study including at least 12 hours of classroom instruction per week.

☐ **RETIRED ($20)** A former Governmental Representative, Corporate or Individual Member who has retired.

New Governmental and Corporate Members will receive a free package of 7 code books. New Individual Members will receive one free code book. Upon receipt of your completed application and payment, you will be contacted by an ICC Member Service Representative regarding your free code package or code book. For more information, please visit www.iccsafe.org/membership or call 1-888-ICC-SAFE (422-7233), ext. 33804.

Please print clearly or type information below:

Name

Name of Jurisdiction, Association, Institute, Company, etc.

Title

Billing Address

City State Zip+4

Street Address for Shipping

City State Zip+4

e-mail

Telephone Number

Tax Exempt Number (If applicable, must attach copy of tax exempt license if claiming an exemption)

**VISA, MC, AMEX or DISCOVER Account Number** **Exp. Date**

Return this application to:
**International Code Council**
Attn: Membership
5360 Workman Mill Road
Whittier, CA 90601-2298

Toll Free: 1-888-ICC-SAFE (1-888-422-7233), ext. 33804
FAX: 562-692-6031 (Los Angeles District Office)
Or, when applying online at **www.iccsafe.org/membership**, please enter **REF 66-05-193**.

If you have any questions about membership in the International Code Council, call 1-888-ICC-SAFE (1-888-422-7233), ext. 33804 and request a Member Services Representative.

*Membership categories and dues subject to change.
Please visit www.iccsafe.org/membership for the most current information.

REF 66-05-193